THE CREATURES UNDERNEATH

THE CREATURES UNDERNEATH

Written and Illustrated by

Jennifer Owings Dewey

Red Crane Books

Santa Fe

FIRST EDITION

Printed in Canada

Design by Jim Mafchir

Library of Congress Cataloging-in-Publication Data

Dewey, Jennifer.
    The creatures underneath / Jennifer Owings Dewey. — 1st ed.
        p. cm.
    ISBN 1-878610-39-2 (cloth) — ISBN 1-878610-48-1 (paper)
    1. Cave fauna — Juvenile literature. 2. Desert fauna — Juvenile
literature.   3. Soil fauna — Juvenile literature.   4. Household
pests — Juvenile literature. [1. Cave animals.   2. Desert animals.
3. Soil animals.   4. Insects.] I. Title.
QL117D45 1993
591.56'4 — dc20   93-33847
                                                              CIP
                                                              AC

Red Crane Books
826 Camino de Monte Rey
Santa Fe, New Mexico 87505

For my grandson Kyle

CONTENTS

MY JOURNEY UNDERGROUND

One hundred twenty feet belowground I felt trapped. My guide, a park ranger, insisted I douse my light and experience the real darkness of a cave. I yearned to keep my light on but did as he bid. I knew I was safe. It was the pure, intense darkness all around that unsettled me.

After what seemed an eternity, probably four minutes, the guide said it was all right to turn my light back on. The instant I did so, my fear passed.

We scrambled and crawled to the entrance of the limestone cave. When we got there I burst out and tore down the hill, grateful to be released from that tomb of darkness.

Curiosity leads us to explore belowground environments. We go as visitors, uninvited guests who want to know more about what lives underneath.

Humans tend to mistrust what we cannot see. A flock of ducks in the air or a housefly banging on a window screen are familiar sights. They do not frighten us.

Later, when I thought about my fear in the cave, I knew it was the darkness that scared me, not the cave itself. I am a stranger to that kind of darkness. A lightless place conceals mysterious animals living invisible lives.

Stories of underground animals are many and complicated. Telling the cycle of what lives below means selecting and choosing only certain animals, this one or that one, and not another, in order to compose a clear picture. Perhaps by describing the life of a mole, earthworm, or bat, the unique nature of animals that live below can be revealed.

DRAGONS, VAMPIRES, AND CREEPY-CRAWLIES

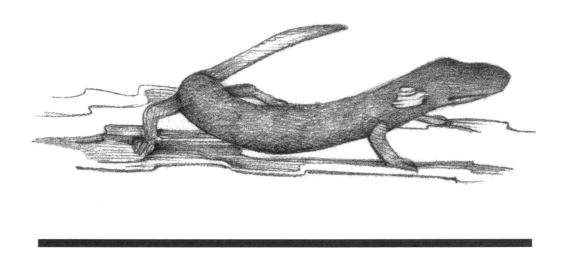

In the seventeenth century a German postmaster went trout fishing in his favorite spring. There the water spilled out of a hole that led to a cave deep inside a mountain. Local legend had it that a dragon lived in the mountain. Each time the dragon shifted position, water surged from the spring.

The postmaster fished for a while. He became annoyed when his line seemed to snag. When he bent over the water to free the line, something caught his eye. It was a tiny dragon. The dragon was eight inches long. It was pale pink and had four feet. Each foot had three toes. The postmaster considered himself an amateur biologist. He gave the dragon a name, calling it Proteus. In Greek mythology, the sea god Proteus was able to change his shape at will.

He returned again and again to the spring. With any luck, he figured he would see more dragons. One day he saw baby dragons, creatures with pinhead-sized eyes that lacked corneas or sockets. The postmaster knew, from looking at the babies, that they came from a dark world, a place where there was no need for eyes.

The postmaster was the first person known to have made note of the blind cave salamander. Dragons are pretend. The blind cave salamander may look like a dragon, but it is not a dragon at all. The blind cave salamander is designed for a life in total darkness, a life underground. It is eyeless and almost colorless. It has fishlike lungs, but it is not a fish.

Many cave animals share these same traits. In a world where no light exists, there is no need for eyes to see with. Where the sun never shines, an animal does not need pigments in its skin to protect it from the damaging ultraviolet rays.

Many cave animals have evolved ways to see without using eyes. Cave fish have special cells on the sides of their bodies that sense movement. These eyeless fish use their skin as eyes to see with.

Cave rats have eyes, extra large ones. Even so, it is their sensitive whiskers that inform them of what is running about in the dark. When cave rats venture into the dark zone of a cave, they leave a trail of urine to follow on the way out.

Cave insects, of which there are countless numbers, use

their antennae, or feelers, to locate a meal and make their way in darkness. Trillions of beetles and their larvae squirm about in guano, or bat droppings. These animals have no idea of light as we experience it.

All I knew the first time I went into a cave was that caves are dark. I knew nothing about what lives and breeds in caves. Limestone cave formations are home to more organisms than other kinds of caves. Water creates tunnels, rooms, and chambers. It seeps and drips through the limestone, carving and etching, a process that takes millions of years.

My own way to understand a cave environment is to divide it into sections, or zones. The first is the entrance, where some light shines each day. Troglozenes, or "cave visitors," frequent the entrances to caves. As the name says, troglozenes come and go from caves and do not stay in the shadows for long. Snakes, mice, and birds are among the animals that enter caves to hunt, sleep, or rest, and then leave.

Beyond the entrance is the twilight zone. Light is dimmer and less dependable here. Troglophiles, or "cave lovers," inhabit this region. Bats and insects compose the biggest troglophile populations. The insects are impossible to count, but people have found ways of counting bats. Using cameras and computers, scientists record numbers of bats as they leave their roosts at dusk. They have found as many as a million bats living in a single cave. Because bears sleep in caves a part of each year, they are considered true troglophiles.

Spiders usually choose the twilight zone for their web building. There are so few flying insects in cave twilight zones that troglophile spiders must go on foot in search of prey.

The blackest depth of a cave is called the dark zone. Troglobites, or "cave dwellers," live in cave dark zones. Troglobites never venture from a life of everlasting gloom, unless by accident or chance. Underground floodwaters, or a surge in a spring like the one in which the postmaster fished, may wash a troglobite to the surface.

Blind cave fish navigate cold cavern streams alongside eyeless white lobsters. Bacteria and algae grow in patches on cave walls and floors, providing a source of food for mites and springtails. These tiny animals are in turn preyed on by larger troglobites, crickets, or salamanders.

Once when I was inside a dark zone, I wondered about my chances of survival if I got trapped. Creatures that are flooded into caves, or blown in, must find a way to adjust or they die. Adjustment involves the passage of time, evolutionary time during which an organism changes shape. Tissue and bone are transformed from one generation to another. Eventually a cave animal is so well designed to life in a cave that it may perish if exposed to a new and different environment.

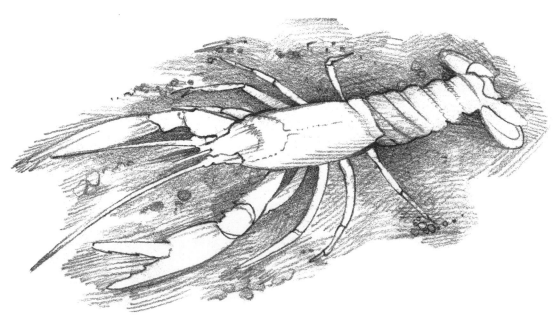

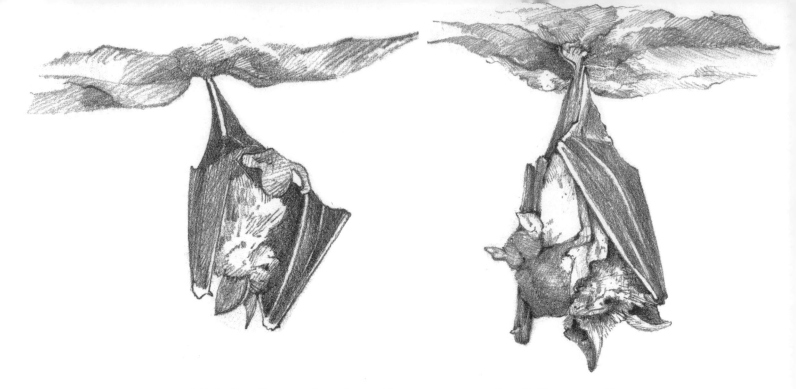

Think about the bat. Bats are troglophiles, or "cave lovers." They are good examples of physical adaptation that allows them to live as they do.

Like other mammals, bats give birth to live young. Their babies, or pups, are born feet first. The mother clings to the cave roof using claws on her hind feet and thumbs. She curls her tail around her belly to catch the baby as it comes out. The bat pup creeps across the mother's stomach and latches onto a nipple. It begins to nurse.

The baby stays with the mother for about a week, holding onto her belly fur by its claws, even when the mother is out flying and feeding.

After a week, the pup is left behind, hanging from the ceiling at the roosting place, while the mother hunts for prey. Many helpless bat pups fall to the cave floor and are eaten by beetles and other insects.

When the mother bat returns at dawn, she identifies her baby by its high-pitched squeak. It is with squeaks and echoes that bats find their way and locate prey. The sounds bats make bounce off objects around them and return as echoes in the animals' ears. In the brain, the echoes are translated into pictures of the physical world and of prey. This method of seeing in the dark is called echolocation.

Bats are wonderful fliers. On the ground, most bats shuffle rather than walk. One bat good at walking is the vampire bat of South America. It walks with the aid of tiny hooks on its elbows.

The name "vampire" is misleading. It suggests sucking blood from the tender white throat of a sleeping maiden on the top floor of a castle. Vampire bats select cattle from which to draw blood, not humans. Vampire bats roost all day and fly out at night to feed. They search for corrals where cattle are gathered. The bats land on the ground and slowly approach a dozing victim. Once the bat is sure its victim is asleep, it climbs up an ankle or flies to the back of an ear. Then it makes an incision with two razor-sharp teeth.

As blood flows, the bat sucks the bright red fluid into its mouth and laps with its long, club-shaped tongue. Vampire bat saliva has an enzyme in it that prevents clotting. More than one bat may feed from a wound. A bat may even return night after night to the same cut, reopening it and drinking more.

Vampire bats are dangerous only because of the diseases they carry. A cut may become infected and need treatment. The incision, and the amount of blood drawn, pass unnoticed by the victim, except in the case of an infection.

There are more insects on earth than any other form of life. The first time I went into a lava cave in the desert, it should not have surprised me to sense the floor moving under my feet.

Insects live and breed in guano, the droppings from bats, which layers the floors of bat caves. The insects exist in such large numbers that the guano itself seems to be in motion. Beetles and their larvae, spiders, mites, cockroaches, and many other insects wriggle and tunnel in the gray, fluffy material. Guano, left behind by generations of bats roosting on the ceiling, covers every surface of a bat cave, resting as a shawl does on a body.

I saw maroon-colored millipedes, their segmented bodies as thick as pencils. Centipedes the shade of vanilla ice cream crept on guano pathways shared with two-inch-long black beetles. The beetles moved as if each one had been struck on the head. Cave atmospheres are without fresh breezes, and the still air encourages slowness.

Most caves are wilderness areas, and no two are exactly alike. One may be home to millions of bats, while another has none. One might be host to a wide assortment of crawling, creeping animals, while another is barren.

I chose to spend time in the desert bat cave because in that cave there were vast populations of all kinds of animals. But it was an effort for me to adjust to the stink of the guano of insect-eating Mexican free-tailed bats. The shed fur of these animals drifted down from the ceiling, mixed with a fine, slow-falling rain of urine. Moths circled lazily in the fuzzy half-light. Daddy-longlegs and cave crickets bunched in colonies on the cave walls, as if life depended on existing in thick piles.

I found a spiderweb ten inches wide, the silky strands laced with water droplets, and I wondered where the water came from. Then I discovered a miniature spring, a seep in the lava. White mosses and pale, hair-thin grasses grew

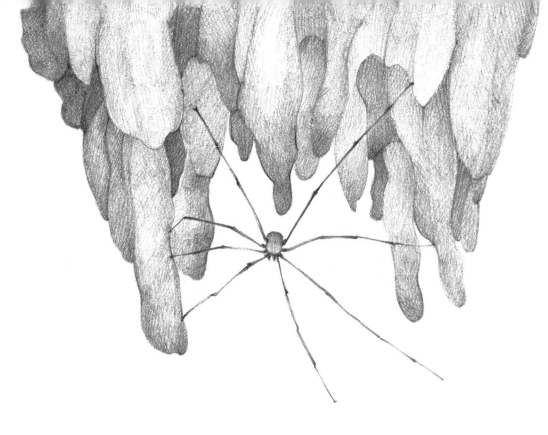

around the opening. A skinny spider, nearly too small to see, had hitched its quarter-sized web over the center of the spring. This spider would never know thirst.

Millions of beetles were creeping along the guano-covered cave floor. The primary species of cave beetle is the dermestid beetle, a half-inch-long, dull-brown animal that feeds on the flesh of fallen bats. These scavengers eat only the meat and leave the bones. Bat skeletons litter cave floors, proof of the harshness of life in such an environment.

In North America there are forty thousand known caves. They are fragile worlds. Heat that radiates from the bodies of human explorers may destroy mineral deposits on a cave wall that have taken decades to form. If you think of cave surfaces as being like fingernails, it is easier to understand what a gash, gouge, or tear means. After an injury, the process of growth must begin all over again.

Despite heat waves and ice ages, cave temperatures, called fossil temperatures, stay constant for millions of years. The perfectly preserved corpse of a saber-toothed tiger, an animal that became extinct ten million years ago, was recently discovered in a cave in Europe. This tiger is surely strong evidence of the absence of change in cave atmospheres.

Entering a cave dark zone means shutting out the real world and all of its light. A flashlight is of little use because cave darkness is so intense. It is astonishing that any organism might find a way to exist in the perpetual midnight belowground.

Some do. Such creatures crawl, creep, slither, cling, or scuttle. By one means or another, they survive and reproduce, in total darkness.

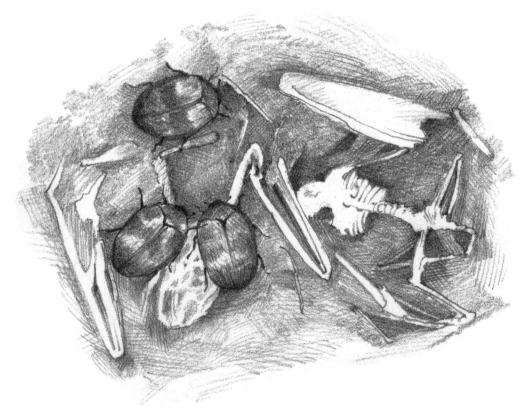

BADGERS,
BURROWING OWLS,
AND
BABY DINOSAURS

The moon rose full in the night sky. Its light bathed the southern New Mexico desert with an eerie glow. My camp was on the sand near where the lava began. I needed soft earth under me to sleep well. I curled deep into my sleeping bag and fell asleep and dreamed. My dreams were about what I had seen that hot July day.

I had explored a lava flow, a sixty-foot-high formation that stretched five miles from edge to edge. Iridescent beetles and black grasshoppers had skittered around my feet. An ebony-scaled rattlesnake had startled me when it suddenly appeared, black-on-black, moving slowly along a bank of rough lava rock. Animals living on lava turn black after several generations. Their fur, scales, feathers, or skin adjust to match the environment.

When morning came, I rolled over and opened my eyes, expecting to catch the first rays of sun streaking the sky. Instead, I was face-to-face with a badger. The badger must

have been watching me for a while. It sat on its haunches staring, as if waiting for me to wake up. I stared back, deciding the best idea was to act as if nothing special was happening.

I knew there were lots of animals living underground in this neighborhood, including prairie dogs and burrowing owls. A coyote pair denned nearby, and of course there was the rattlesnake I had seen earlier. But the badger was a bonus and an unexpected surprise.

The badger kept its head still and watched me, its dark, deep-set eyes fixed in a steady gaze. I got the impression the animal was nearsighted. I was tempted to run my fingers through its thick, dusty pelt. But I knew better. Badgers are powerful and aggressive animals. They have sharp teeth and nasty tempers when cornered.

The badger took a step back and sniffed the air with its chin slightly lifted. Badgers have better noses than eyes. When they hunt, they rely more on their noses than on their vision.

Finally the badger moved off on silent, padded feet. I wanted to find its den. If I managed to keep far enough back, maybe the badger would forget me and I would be able to observe it. It took three days to find the den. I perched sixty yards away and kept as still as I knew how.

There were two adults, which I took to be a mated pair. I counted three babies, each the size of a flattened soccer ball. The young ones were sheer bundles of raw energy. When their parents let them out of the den, they snuffled and sniffed around, rolling on the ground and loading their coats with dust. They explored a world new to them, and examined each other, using their sensitive noses.

Parents keep track of their brood by herding them close to den openings. The little ones learn by imitating their elders. Digging is a favorite activity, a skill these animals are born with. Badgers work all the time to keep their tunnels and subterranean rooms clean. There is always a lot of digging and clearing out after a summer rain.

Morning is the time for housekeeping. Grass bedding from the day before is taken away and replaced with fresh bedding. Entrances and tunnels are redug and packed anew. Badgers do not move quickly unless under attack. Still, they get a lot done in a short amount of time.

While the babies remain at home, the parents hunt. The adults sniff out voles, lizards, prairie dogs, and an occasional burrowing owl. Even though I know that for everything living something must die, the first time I saw badgers returning to their den with slain burrowing owls, it gave me a start.

One searingly hot afternoon I crossed paths with a

seven-inch-tall owl the color of dust. The bird bounced and skittered over the ground at top speed. I decided to follow.

The owl moved like a firecracker out of control, hanging right and left turns around cactus and creosote, the desert shrub that smells like sugar and oil after a rain. It swiveled on spindly legs until it came to a mound of earth surrounding a nest hole. In a single gesture it spun around to face me, letting loose with a throaty hiss and staring with yellow eyes. I was a trespasser. I was being told to clear off.

I had come upon a burrowing owl colony, home to about thirty owls. These birds live in the ground and are difficult to count. They tend to look alike. I was never sure if I was seeing the same owl over and over, or a new one each time. A grassy ridge, thick with Apache plume, provided a perfect lookout spot for me to sit on. In no time at all, the owls had forgotten me.

Owls devote hours to grooming. Parasites, such as fleas and mites, are a serious problem for belowground animals. One method for an owl to rid itself of these is for one owl to draw the feathers of another owl through its beak. A second technique is to roll in the dirt and shake furiously afterward.

Burrowing owls are squatters. Although their scientific name, Athene cunicularia, means "little miner," these birds depend on other animals to create their living space. Prairie dogs, the original diggers of the owl holes, were still present. Members of the ground squirrel family, prairie dogs outnumbered the owls ten to one. Their town stretched for more than three acres.

Prairie dogs and burrowing owls are diurnal, meaning they are active in the daytime. These animals not only share quarters in the ground, they have common enemies.

Another day, a three-and-a-half-foot-long western diamondback rattlesnake slid by my observation post. The snake's scales were a soft, subtle, grayish green blended with white and tan. Powerful reptilian muscles labored in perfect harmony to pull the snake across the sun-drenched earth. It slithered close to the mounds and nest holes, flicking a forked tongue as it went. Every prairie dog and owl within one hundred yards dove into the ground the instant the snake came in view. I was alone with it.

My fear passed and was quickly replaced with admiration. The animal was a reptile, yes, but it was beautiful and efficient. The snake spent a lot of time sniffing and examining the openings to burrows, but it must not have been

hungry that day. After a while it slithered off, perfectly matching the sage it crept through. A slight track on the sand was all that remained as evidence it had been there.

Rattlesnakes often den-up in prairie dog holes. The close association between prairie dogs, rattlesnakes, and burrowing owls creates a tense atmosphere under the ground. Snakes prey on all of their neighbors.

Hawks, eagles, and falcons drop out of the sky overhead, while coyotes, badgers, and foxes slip silently into a tunnel or a burrow to capture a fresh meal.

Burrowing owls have found a way to fool the enemy. Where these owls live close to cattle or horses, their dung contributes to the owls' survival. Female owls incubate their eggs and keep the hatchlings warm inside the burrow belowground while the males serve as guards and defenders. In areas where dung is available, males collect it in their beaks and drop clumps of it around a nest hole opening. A passing fox, badger, or coyote is unable to catch the scent of owl because the aroma of dung conceals it.

Many life-threatening events make life touch-and-go for animals living belowground in the desert. Summer thunderstorms bring so much rain at once, the animals have nowhere to go but into dens and burrows. Animals are flooded out and must seek high ground until the water is drained off or absorbed into the earth.

    While rain makes survival difficult for some animals, for the survival of others it is essential. A four-foot-wide waterhole one hundred yards from my camp was dry until the summer rains began. Then it was transformed into a densely populated breeding pool for spadefoot toads.

These toads are named for their large back feet and toes, which they use to dig themselves into the ground during dry times. The toads are protected inside their underground tombs by a jelly secreted, or released, from their skin. The jelly creates an envelope of moisture.

When the rains come, the toads dig out, take up residence in a puddle, and begin to croak. The croaking draws the attention of other toads in puddles elsewhere in the desert. Males and females find each other. Eggs are fertilized, tadpoles hatch, and in a matter of a few short weeks the young are able to dig into the earth as their parents do.

Spadefoot toads require puddles to mate and lay eggs. They may stay in hibernation for days, weeks, or even months, depending on the presence or absence of water from the sky.

In most environments on earth, the coming of night brings change. Nocturnal animals, those that hunt under cover of darkness, come out of dens, burrows, and nest chambers. They prowl around in search of prey and mates.

A coyote pair had a den less than a mile from my camp. These song dogs of the West mate for life. The arrival of pups is a signal for the male to move out and den-up alone until the pups are more than half grown. During this time, the male and female meet at dusk to hunt together. Night after night I heard their calls, the yipping and yapping of one to another. Their calls were a way of saying, "I am here, where are you?"

I used to believe that ghosts had parties in the darkness of night. The coyote calls were like the voices of ghosts say-

ing when the party would begin.

I saw the pair several times but only glimpsed the pups once. One view I got of the pair came as they returned from a successful night of hunting. I could tell they had caught something. Their bellies were fuller than usual.

The moon, important all the time, is especially useful in the desert and when walking across lava. Its gentle light helps define where you are going and where you have been.

One night I ventured away from camp. I saw a kangaroo rat in mortal combat with a huge grasshopper. It was a no-win situation for the insect. I also saw centipedes and dodged scorpions. I carried a stick in case a rattlesnake might be meandering around in the dark. Ridges and crests of lava rose in front of me. Thirty-foot canyons fell away below me. I pitched and plunged over the lava and soon got tired.

Sandpits, low places where windblown dust and grit collect to create enough soil to support the roots of grasses and cactus, dot the lava. Creosote grows in the low places.

I chose a moonstruck sandpit to rest in. The minute I sat down, I was aware of being watched. I gazed along the top rim of the lava wall encircling the hole, and then I saw them — two hairy tarantulas backlit by moonglow. They looked at me, eight eyes on each furry face gleaming like headlights on a couple of high-speed trains.

I waited to see what would happen next. The tarantulas turned to face each other. One raised its two front legs into an arched position, and the other spider backed away.

In a flash the two were joined, a tight clasp of thick-limbed bodies in a fierce embrace.

I thought of movies I had seen where a loving couple danced by moonlight on desert sand. These cinema scenes ended with a romantic kiss and vows of eternal love. This would not be the destiny of the tarantulas.

After mating, which is what the spiders were doing, the male would leave. Possibly he would dart away as fast as he could. Female tarantulas have been known to make a meal of a male, if hunger drives them.

I climbed out of the sandpit with care, not wanting to squash the pair of tarantulas engaged in the activity of making more of their kind.

The tarantulas I saw are common to deserts in the southwestern part of the United States. They have red, hairy knees and grow to be four or five inches wide. Their days are spent in burrows in the ground. These spiders line their burrows with silk.

At night tarantulas come out of the ground and seek prey or attempt to find mates. If rains fail to come, tarantulas may become dehydrated and go in search of wetter places to dig burrows.

Like vampire bats, tarantulas are misunderstood. Perhaps they have too many legs or too much hair. Their venom may kill a small rodent, but it is harmless to humans.

My desert sojourn exposed me to many new and hard-to-see animals. Especially nice was the discovery of herds of baby dinosaurs.

Desert days, like any others, blend after a while. The memory of baby dinosaurs scampering on the ground keeps those days fresh, no matter how much time passes.

As a child, I was aware of animals in the natural world. There were skunk tracks in the mud, snakes in the rasp-

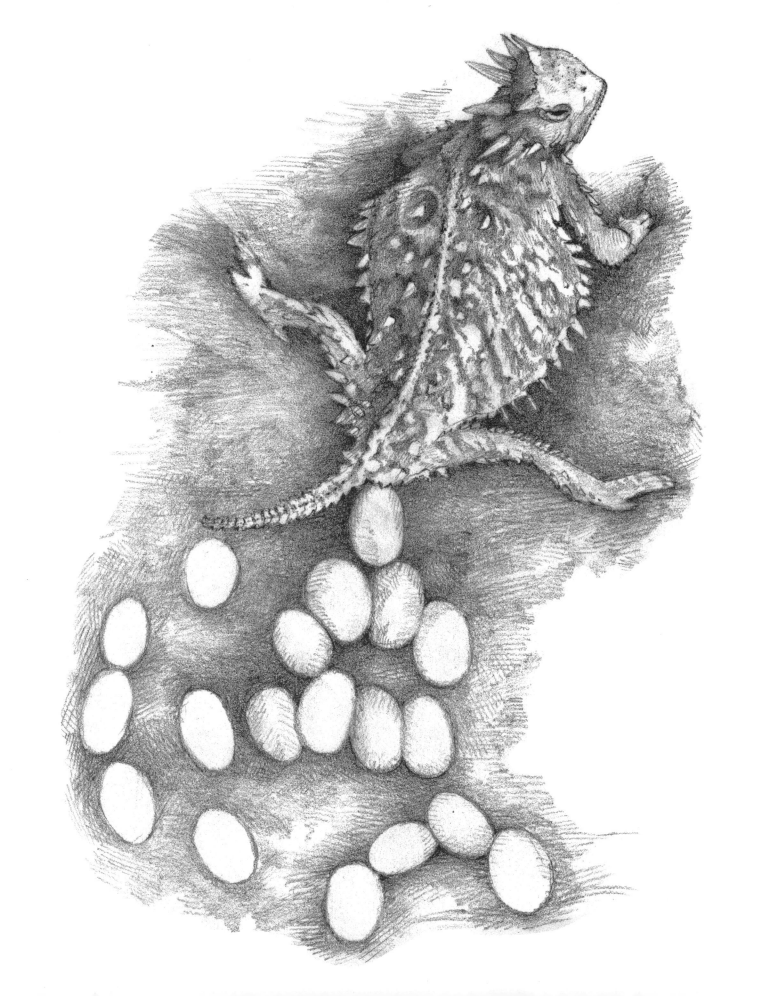

berry bushes, and rabbits in the alfalfa. But nothing measured up to finding a baby dinosaur by the front gate. Martha, my best friend, was with me when I saw it. Like me, she had books at home with pictures showing the pentaceratops and the brontosaurus. If we knew anything at nine, it was what a baby dinosaur looked like.

We captured the animal and took turns holding it. We rubbed the cream-colored belly and were alarmed by the red fluid oozing from the corners of its tiny black eyes. It was three inches long and the color of the ground, a mixture of brown, white, gray, and black. We were puzzled by how docile it was, making no effort to bite.

Years later I knew we'd held a horned lizard. With spines and horns, bony fringes, and spikes on its tail, it is no wonder the animal was, in our minds, a baby dinosaur.

Horned lizards eat ants. Near my desert camp there was a huge ant nest. The lizards were plentiful in the neighborhood because of this ample food supply.

Flat-bodied horned lizards are masters of camouflage. A horned lizard living in one desert may be colored very little like another living nearby. Their coloring matches the ground they travel over.

My lizards were yellow, brown, tan, and black. I searched to see if horned lizards lived on the lava, but I found none. It would have been interesting to see their scales and horns the color of black lava rock.

To a horned lizard, home is a hole in the ground. They are active in the daylight hours, so long as the sun is hot. The sun warms their bodies and allows them to scamper about and feed on ants.

Millions of years ago, creatures similar to lizards lived in the sea. None lived on land. In time a migration took place, and these sea-going animals evolved into land dwellers. The Age of Dinosaurs came, a time when dinosaurs dominated all life on the earth.

The lizards we see today may in some respects look like miniature dinosaurs, but they are not. Like the dinosaurs, they are successful animals because their skin reflects heat and resists the loss of moisture. Their eggs have leathery skins and can be left alone after being deposited. The pancake shape of a horned lizard is another survival strategy, along with horns and plates. A wide body allows for more than a few number of eggs to be carried by the female. A greater number of young have a chance to grow up.

The red fluid that upset my friend Martha and I was blood. The blood comes from glands close to the horned lizard's eyes, glands like our own tear glands. When a horned lizard is attacked or frightened, it squirts blood from these glands to scare the enemy away.

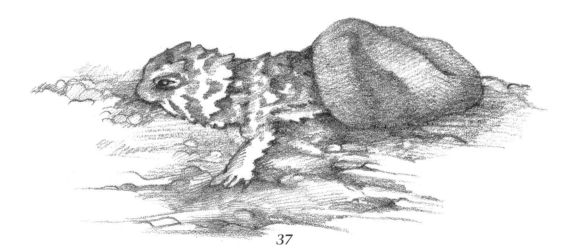

On the day I scooped up my bedroll and other camp possessions, I knew it was the horned lizards I would miss the most. Luckily for me there is a colony of these below-ground animals in my backyard in town.

The lizards are so matched to their background that I fear I might step on one. I am careful because it would be a shame to hurt a baby dinosaur.

WIGGLERS
AND
BUILDERS

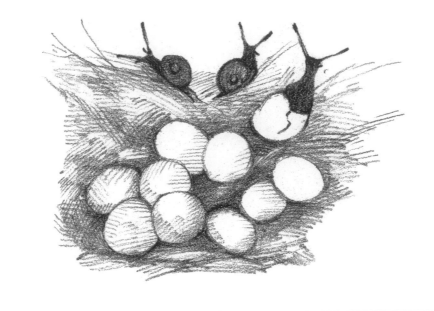

At age eleven I decided to become a worm farmer. It was a profession suited to a child, a way for a kid to make extra money.

I lived on a pig farm. One of the farmhands helped me build the coffinlike boxes needed to house the worms. This part went well. Next came the soil preparation, combining water with fertilizer to an exact consistency my "livestock" would thrive on. Then came the filling of the boxes, which seemed to take forever.

My parents loaned me the money to buy worms from a man in town who had his own worm operation. It was with high expectations and a thumping heart that I dumped the slippery, sliding creatures into their homes.

The third morning of this project I went out as usual, before dawn and breakfast, to see about my worms. Every single one had vanished. I wept and blamed myself for creating a home the worms hated so much they all left. Later I

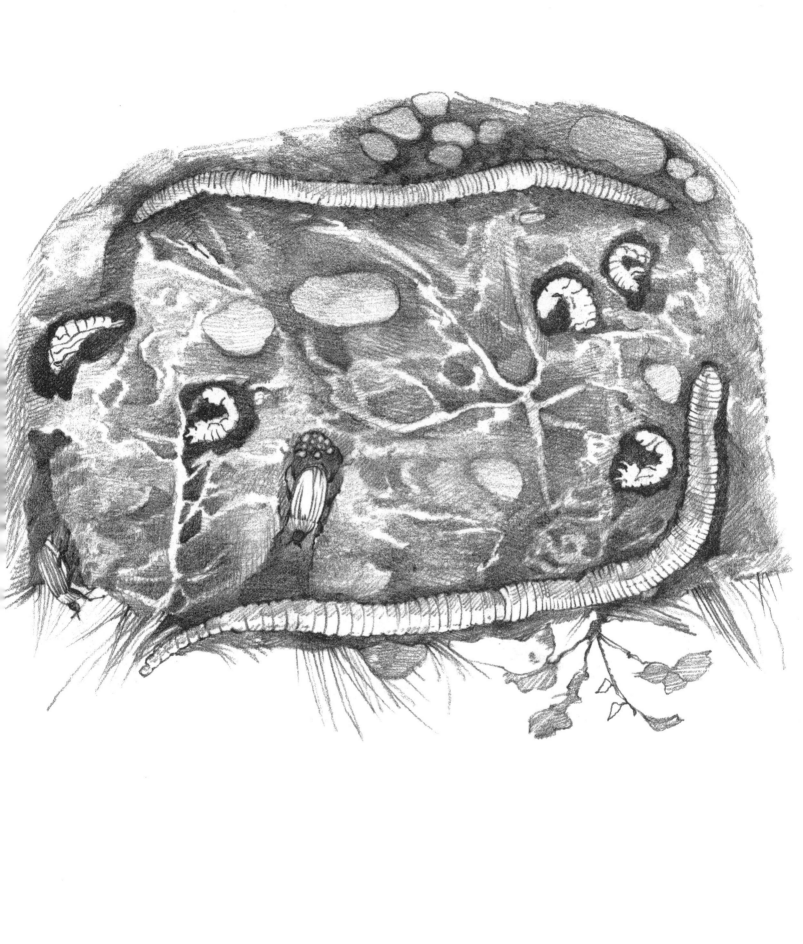

learned that although worms have tiny brains and slimy skins, they are sensitive animals.

I will never know for sure why my worms left when I was eleven and trying to make a good home for them. I do know the right blend of soil, water, air, and bacteria makes a difference. If one of these is out of balance with another, a worm farmer's efforts may be wasted.

The mix of ingredients attracts animals and enables them to set up housekeeping. Some soils are crawling with life, while others have not a single worm or mite.

Levels of moisture are important to mole crickets, the growth of insect eggs, and the larvae of beetles. Invertebrates,

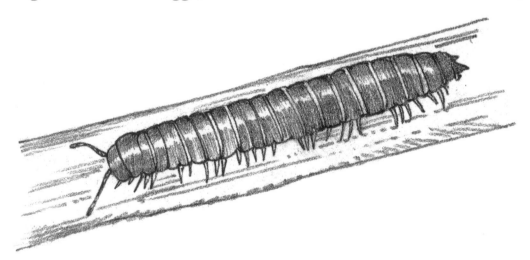

or animals without backbones, make up most of what lives belowground. These animals have shapes, forms, and lifestyles that require moisture in the ground for the sake of survival. If they dry out, they die.

I have witnessed migrations of millipedes in the desert, hundreds of thousands at once creeping in search of damp

soil when there has been a drought. Self-preservation brings these many-legged animals out of their protective burrows without any assurance they will find a better place to live.

Springtails live in soil, millions to the cubic inch. These shrimp-shaped animals can leap many times their own length, a skill that is useful when soils dry out and there is a need to move to wetter ground.

Moist earth is essential to the life of a slug or a snail. These animals are gastropods, a word that means "belly-foot." Their flat undersides are a muscle that moves them along like a foot.

Slugs are snails without shells. They secrete, or ooze, a

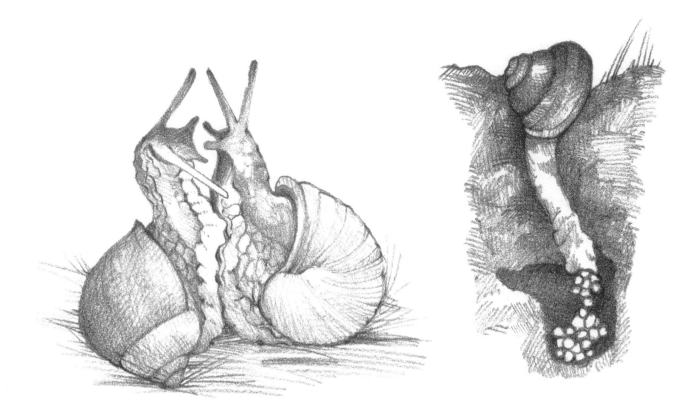

slimy fluid that covers their bodies and allows them to creep around. If they dry out, they cannot move and they die.

Dig a hole with your hands and look at the earth sifting through your fingers. Depending on where you dig it may be bone dry, sticky-wet, crumbly, or sandy. Some form of organism probably lives in your earth, even if it is more sand than fertile soil. Forest floors, cultivated fields, and meadows in the mountains are rich with populations of belowground animals. Ages of decay and bacteria feeding on roots, stems, seeds, and what drops from trees and bushes nourish many forms of life.

Soil dwellers are often active more hours of each day than creatures living on the surface. Earthworms hardly ever rest. They wriggle through the earth, chewing, digging, digesting, and excreting soil. Worm leftovers are called

44

"casts." These are little bundles of worm juices and dirt. An earthworm's soft, segmented body, made mostly of muscle, shifts its shape from fat to thin when the need arises. Some soils are filled with tiny stones or pebbles, and a worm must "shrink" to get around these.

Earthworms are hermaphrodites. The word means they are male and female at once. Even so, they must join to make eggs. They lie side by side, their bodies connected by a sticky tube. After this, both worms go off and lay eggs. Their eggs are enclosed in cocoons about the size of a large pinhead. Baby worms hatch and are immediately able to do whatever adult worms do.

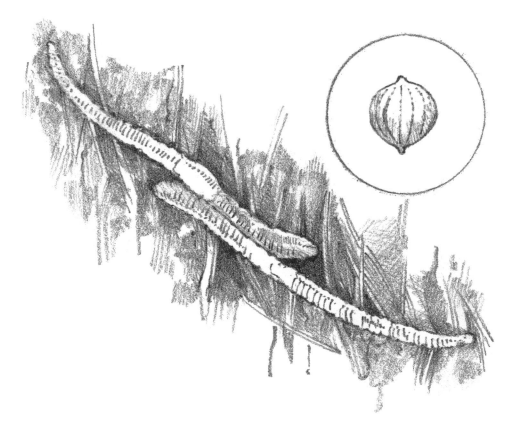

Most invertebrates do not tend their young. They find safe places for their eggs or build sacs or cocoons around them to protect the next generation. Usually, that is all they do.

The beetle Nicrophorus, or "gravedigger," has a method of raising young that is unique among insects. Unlike others of their sort, burying beetles invest a lot of time and energy in the care and feeding of their offspring.

Burying beetles have antennae that serve as noses. By tracking down the smell of a dead animal, a mouse or a bird, these beetles begin the process of mating and child rearing.

The corpse is walked to a site chosen earlier. Male and female crawl under the dead animal and inch it along. Next is the burial.

The animal is buried three or four inches below the surface. The beetle pair work together to strip the body of its fur or feathers and coat it with secretions from their back ends. The fluids embalm the body and help prevent rot.

Then the beetle pair mate. Two days later, the female lays twenty or thirty eggs in a small depression on top of the corpse. The larvae hatch and within days feed themselves. The whole family uses the dead animal as a food source. If there are too many young for the amount of dead flesh, the parent beetles will eat some of them so those remaining can survive. What is left in the tomb is nothing more than a few

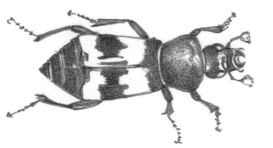

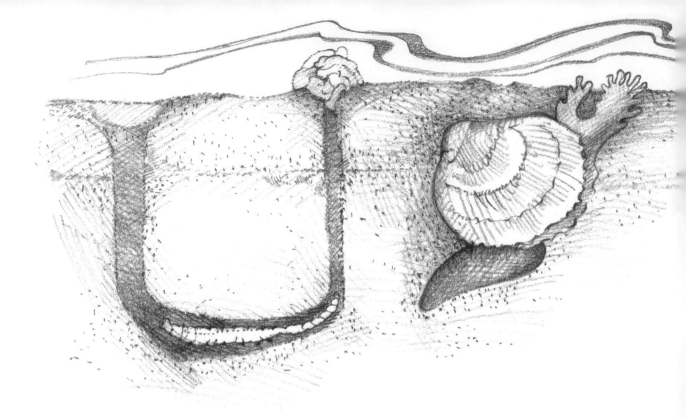

of the dead animal's larger bones.

In about nine days the male beetle leaves; his job is complete. The female goes soon after. The larvae are ready to squeeze and wiggle themselves into the ground next to the burial chamber. Here they will pupate, or change into adult burying beetles.

Sand has different animals living in it than does rich, thick soil. In African deserts where heat is intense, many animals, mostly lizards and spiders, spend their day in the ground and emerge only after the sun has set.

The ocean has tides where the waters rise and fall. Seawater comes in and goes out in a pattern. Clams and lugworms, both animals that burrow down, feed off organisms washed in on high tides.

A clam secures itself in the sand by a single foot, a muscle able to hold it in place even during storms. When the

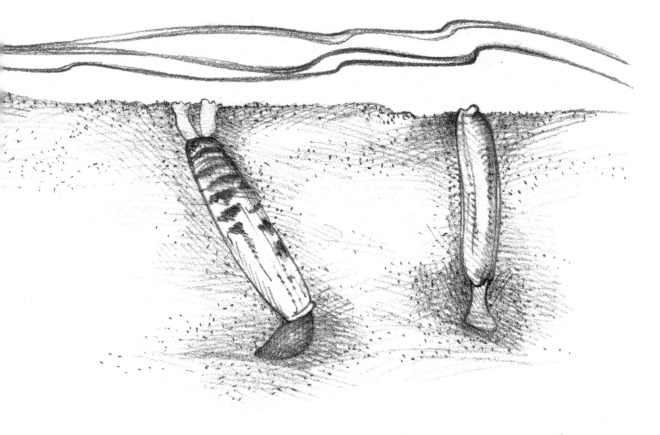

tide comes in, the clam takes edible particles from the water.

Lugworms live in U-shaped burrows in the sand. Lugworms feed as clams do. At night lugworms leave their holes and creep around. Their excretions, or "casts," can be seen on the sand near the openings to their homes.

As a dryland person who did not see the ocean until the age of fourteen, the environment does not feel natural to me. When I walked on a California beach and kicked up a strand of kelp, or seaweed, I was astonished to see millions of sand fleas. There were so many of these leaping, bounding animals they created a cloud. The sight impressed me. Just as there are numberless stars in the heavens, there are countless fleas in the sand. Sand fleas try to stay out of the sun. The light is deadly for them. Some of their enemies are crabs, seabirds, and birds that live on the sandy shore, feeding off whatever insects they can find.

Predators are thick and fast for animals below as well as for those on the surface. The main enemy of the earthworm is the mole. Moles rarely rest from digging. By scooping with their large front feet, they move tons of earth all over the planet. With poor vision but tireless determination, moles, like earthworms, make soil healthier by creating passages for air.

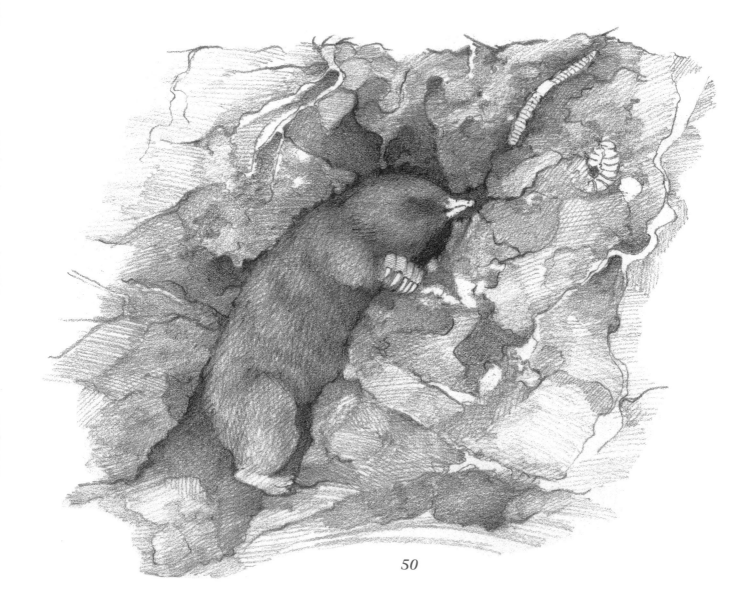

When a worm falls into a mole tunnel, the mole is after it in an instant. The mole bites down in such a way the worm is paralyzed and not killed outright. Moles store worm victims in rooms under the ground. A mole with a worm-filled pantry has an edge against hunger when winter comes.

As a child, I crawled across the ground on my stomach in an effort to view the world as a mole would. My friend Martha dangled worms before my eyes so I might imagine what it would be like to capture a delicious wormy meal. But none of this aroused my interest in worms the way our science teacher did.

Mr. Thomas was tall, well over six feet. One morning he came into the classroom and told us to be seated because he had something special to show us. He often brought live animals into our classroom: black widows, June bugs, praying mantids, and scorpions. He never ate any of these.

But on this day Mr. Thomas opened his mouth and dropped in a three-inch-long earthworm. He gulped once as the worm went down his long, skinny throat. Our class watched in silence.

Many of us hated meat, unless it was a greasy hamburger. Mr. Thomas was aware of this. His lesson on the importance of protein was unforgettable to our third grade class. It seems these wiggly, damp creatures are a prime source of protein.

In hard times I will know what to do: serve up a platter of earthworms and save my family from starvation.

There are places in the world where people enjoy the taste of a roasted termite queen or a fat grasshopper dipped in hot oil. I confess I am not able to eat these things.

Termites are world-class builders. Along with the ancients who built the pyramids, termites win the prize for creating amazing structures. Termite nests and mounds endure for centuries.

Like many species of bees and wasps, termites are social animals. Their colonies number in the hundreds of thousands, individual insects working in harmony to create a whole.

With saliva and their own droppings, termites mix and chew the earth they live in to make a substance they build with. On the surface are mounds, the part of a nest exposed

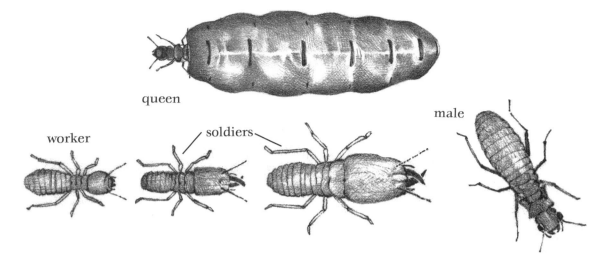

queen

worker

soldiers

male

to sunlight and air. The interiors of mounds are designed to catch and hold the warmth of the sun's rays. Breeding chambers for larvae are positioned on the sunny side of the mound. In wet, tropical climates termites construct roofs over the tops of the mounds to prevent water damage. These umbrellalike formations are made of the same material as the rest of the mound.

Belowground, in a termite nest, are the real workings of a colony—the storage rooms and culture chambers. Termites feed off fungus growing on rotting vegetation they bring into the nest and stash in special rooms.

In a termite city, every termite has its own job to do. This includes the queen. She lives in her royal chamber and exists only to make eggs, eighty thousand in a week's time. The king termite stays close to his queen. His purpose in life is to breed with her and make more termites.

Termite job descriptions consist of workers and soldiers. The workers build, the soldiers guard. These urban insects spend most of their lives in darkness below the ground. It is rare that they rise to the surface and expose themselves to light. Termites come in several shapes and forms. The kind that get into houses and chew on walls, floors, and ceilings, are called "pests."

I once went to my closet to pull out the manuscript of a book I had written. I wished to make a gift of the manuscript to my local library. The manuscript, and the box it was in,

had been reduced to a brownish fluff. Under the box I saw wiggly, white animals trying desperately to get away. Over weeks and months of time, termites had devoured a number of manuscripts in my closet. When I thought it over, I became less upset. Everything living has a right to exist. It was my own fault. I had left the perfect store of food for termites, and they did what termites do: they ate it.

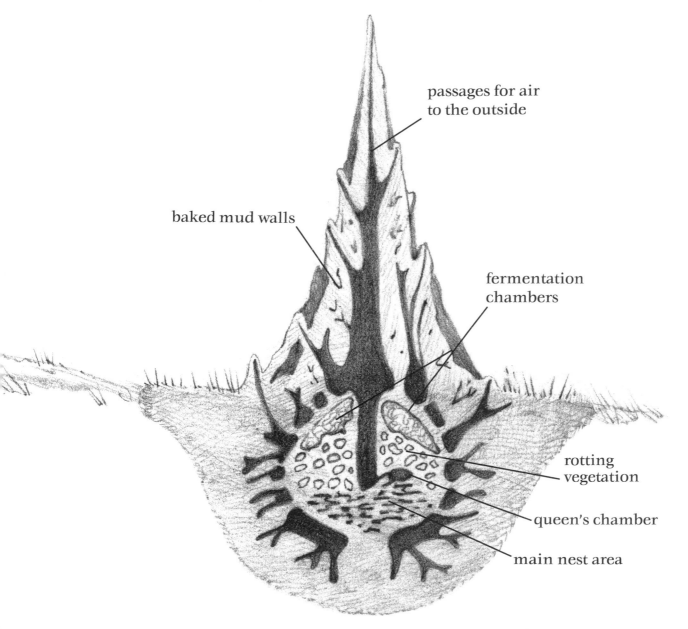

passages for air
to the outside

baked mud walls

fermentation
chambers

rotting
vegetation

queen's chamber

main nest area

CREATURES
UNDER
THE RUG

Houses are environments too. Humans happen to live in them on a regular basis. Unlike a forest or a mountaintop, where animals come and go without much interruption, house-holds present special challenges for nonhuman occupants.

A tiny child of two or three may spend hours closely examining an insect crossing the floor. The child will prob-ably hunch down to the level of the bug and stare, fascinated by its leg action, the shape and form of something entirely new and strange.

Adults tend to squash insects the minute they see them. Or they rush for a jar so they can remove the animal from their house.

I remember a time when I had a new baby in the house. Often I was up feeding the baby in the dark. On one occasion I walked into the kitchen, babe in arms, and reached out to switch on the light. Less than an inch from the tips of my fingers was a four-inch-long centipede clinging to the wall.

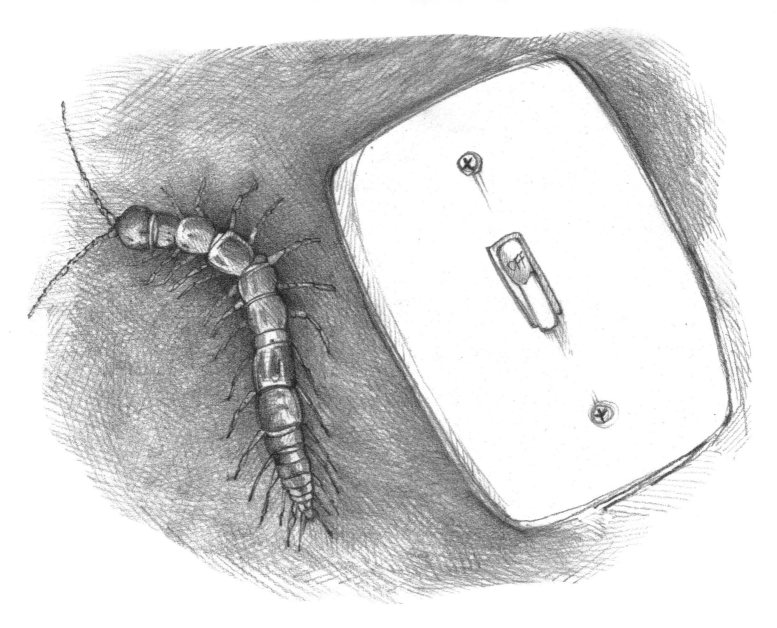

My body whirled around, arms squeezing the infant girl, before my mind took over. Then I realized centipedes of this kind do not bite humans. They go in search of leaf litter and other decaying vegetation collected and caught under doormats and front steps.

My instinct was correct; a mother protects her child. Still, there are things to know about what travels around under a rug or the floorboards of a house.

Insects and spiders seek refuge in houses. Many are too small to see without the aid of a microscope. Most people do not consider or are unaware of the populations of microbes in their kitchen sinks. Houses are crawling with creatures passing before our eyes, and we never know it.

Recently I walked barefoot across my floor in the dark. I stepped on a two-inch-long ground beetle. It survived without injury. When I carried it outside, I advised it not to come back. I do not object to ground beetles, unless they are sharing space with me. Our ways of life do not agree.

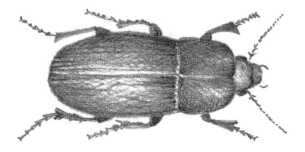

Those insects and spiders successfully sharing houses with human beings have probably found ways to remain hidden. A flea can live in bliss on the hide of the family dog. Others seek dark nooks and crannies where human eyes cannot pry. Basements and attics are legendary hideouts for daddy-longlegs, wandering spiders, and crickets. The presence of these animals usually does no harm, but black widows can present problems for people. They often migrate into houses when cold weather comes. Black widows are among the world's toughest spiders and they have potent, dangerous venom. A friend of mine was bitten once. She became ill, even though she was a sturdy, well-fed teenager. The area of the bite became a nasty ulcer that needed to be removed by a surgeon.

Spiders are unique animals. They trap prey using silk that comes out of their bodies. They build webs and drag lines with this same material. Spiders can go for months without food or water and then revive when a meal is caught.

Other animals likely to be found hidden in houses are millipedes and centipedes. Both seek damp places, such as back steps where leaves and grass cuttings collect. In North America there are no venomous centipedes, and so there is no danger from a potential bite. Millipedes and centipedes feed on decaying vegetable matter and will occasionally eat the eggs or larvae of other insects.

If the wood around windows and doors becomes damp and begins to rot, wood lice may take up residence. Wood lice have eight legs and plates on their backs that overlap, as roof tiles do. Female wood lice have pouches on their sides where they deposit eggs. Five weeks later, baby wood lice hatch out and scatter.

There may be seasonal infestations in houses. Ants are often around in dry, hot times. Seeking water, ants will leave their underground nests and form long columns. They will march until they find a water source. They collect droplets of moisture on their bodies and carry these back to their dried-out nests.

One creature tough to get rid of, once it moves into your house, is the cockroach. These animals are "living fossils," which means they are almost unchanged in physical shape and habits from the time they first appeared on earth millions of years ago. Their bodies resist poisonous chemicals by gradually becoming immune to the effects. This makes it nearly impossible to drive cockroaches out of a house once they have settled in and started breeding.

All structures are host to organisms of some kind, from mites, fleas, and ants to a million other kinds of crawling things. Where you live in the world plays a role in what moves in with you. The southwestern United States is home to beetles with iridescent wings, hairy tarantulas, and an animal called a whip scorpion.

When I was twelve, I met my first whip scorpion. I had been wading in the irrigation ditch, where the water was muddy and smelly. When I got back to the house, I went to the bathroom to run water in the tub for a bath. There, starkly black on the tub's white enamel, was something I'd never seen or even imagined. The animal was about three inches long with a segmented body, crablike legs, and a tail with a stinger on it. Before thinking in a rational way, I grabbed a broom and beat the creature to a pulp.

I learned later that I had demolished a harmless bug, a whip scorpion. They are not venomous. They do not sting.

They have glands on their back ends that contain an acid that smells like vinegar.

A regional name for the animal I killed is vinegaroon. The name comes from the fact that the animal sprays the vinegar-scented acid into the air when it is attacked. The spray is harmless to humans, although it may stun or drive off a bird or a mouse. A vinegaroon uses its spray as a defense, rather than as a way to kill prey.

I did not see another vinegaroon for about three years, but I knew they were there, somewhere, hidden behind the walls, or perhaps under the floor.

Many varieties of bugs, spiders, and insects cohabit with humans. Even so, it is an unlikely choice for most of these animals to make. Just as you or I would be reluctant to live in a bat cave, a mole tunnel, or an ant nest, these creatures prefer their own worlds to ours.

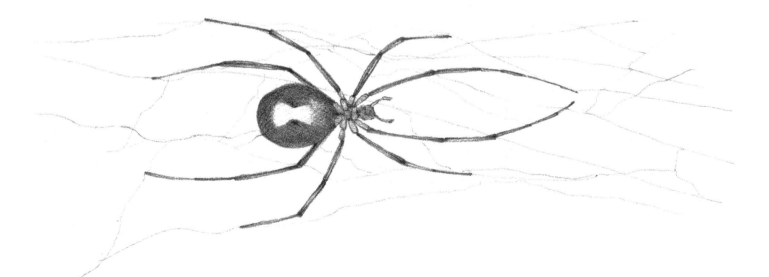

CONCLUSION

As long as humans exist on planet earth it will probably be true that what creeps, crawls, slithers, or skitters below the ground will seem alien to us. This alienation might be a kind of protection for us as well as for them.

When I held my baby daughter and got scared at the sight of a large centipede on the wall, I did what my maternal instinct said to do, I guarded my baby from harm. Any insect living in someone's house will try to protect itself by running or hiding. The experience of seeing the centipede led me to think over how I feel about animals with too few legs, or too many, or those that hide out of sight. In time I realized my own life path might cross a beetle's, and each of us could go on from there, with no harm done.

We may not trust what we cannot see. Still, the answer to the question of trust may be as simple as saying, "live and let live."

GLOSSARY

Age of Dinosaurs   During the Mesozoic era, or the interval of "middle life," reptiles, especially dinosaurs, were dominant on land. For this reason the Mesozoic is often described as the Age of Dinosaurs.

Apache plume   A shrub common to the southwestern regions of North America and Mexico, Apache plume (Fallugia paradoxa) grows at altitudes above three thousand feet. It is named for many fuzzy flower clusters that resemble Indian feather war bonnets. A member of the rose family, Apache plume reaches a height of between three and four feet. The shrub has slender, whitish branches that become shaggy as they age.

cornea   A transparent tissue that forms the outer coat of the eyeball.

drag line   A silken line released by some spiders to control their leaps in space or to help them find their way from one place to another.

embalm   To treat a dead body with chemicals to keep it from decaying. Embalming preserves dead flesh.

enzyme   A proteinlike substance made in plant and animal cells that acts as a trigger for a chemical reaction.

evolution   The gradual change in formation or growth of a population from generation to generation. It is thought that one process that brings about the change is natural selection. Charles Darwin was the first to observe and document natural selection among groups or organisms.

fungus   Any of a large group of molds, mildews, or rusts.

ice ages   Periods of time when vast masses of ice form on the earth's surface.

insect   An animal with a segmented body composed of three parts: head, thorax, and abdomen. Insects are invertebrates, which means they have no backbones. Insects wear their skeletons on the outside of their bodies. They normally have a pair of wings attached to the thorax. Insects usually have two sets of jaws, two kinds of eyes, simple or compound, and one pair of antennae, or feelers.

iridescent   The display of rainbow colors created by refraction of light when an animal shifts its position in the sunlight.

larvae   The young of any invertebrate animal that undergoes change as it becomes an adult. Many insects experience a larval form. The tadpole is the larva of a frog, the caterpillar is the larva of a butterfly.

lava   The melted rock that issues from a volcano. Lava is also this same molten material after it cools off and hardens.

microbe   Any microscopic or extremely small organism.

parasite   A plant or animal that lives on or in another organism in order to feed from that host. Parasites usually do harm to the organisms off which they feed.

pigment   A substance in the cells and tissues of plants or animals that colors them.

REFERENCES

Bosted, Ann, and Peter Bosted. "Big Show Goes On Below: A Story on Cave Animals." *Ranger Rick* (April 1991): 29.

Brodie, Edmund D., Jr. *Venomous Animals*. Racine, Wis.: Western Publishing Company, Inc., 1989.

Burton, Maurice. *Cold Blooded Animals*. World of Science Series. New York: Facts on File, Inc., 1985.

Chorlton, Windsor, with staff editors. *Ice Ages*. Planet Earth Series. New York: Time-Life, Inc., 1983.

Churchman, Deborah. "Rocky Roach Tells All." *Ranger Rick* (September 1993): 4.

Fenton, M. Brock. *Just Bats*. Toronto: University of Toronto Press, 1983.

Geluso, Kenneth N., J. Scott Altenback, and Ronald S. Kerbo. *Bats of Carlsbad Caverns*. Carlsbad, N.M.: Carlsbad Caverns Natural History Association, 1987.

Hill, John E., and James D. Smith. *Bats, a Natural History*. Austin: University of Texas Press, 1984.

Jackson, Donald Dale, with staff editors. *Underground Worlds*. Planet Earth Series. New York: Time-Life, Inc., 1982.

Jones, Dick. *Spider: the Story of a Predator and Its Prey*. New York: Facts on File, Inc., 1986.

Kirchner, Rich. "Owl Pals." *Ranger Rick* (April 1993): 42.

O'Toole, Christopher, ed. *The Encyclopedia of Insects*. New York: Facts on File, Inc., 1986.

Ruffault, Charlotte. *Animals Underground*. Ossining, N.Y.: Young Discovery Library, 1988.

Schnieper, Claudia. *Amazing Spiders*. Nature Watch Books. Minneapolis: Carolrhoda Books, Inc., 1989.

Sherbrooke, Wade C. *Horned Lizards, Unique Reptiles of Western North America*. Edited by Earl Jackson. Tucson: Southwest Parks and Monuments Association, 1981.

Tinbergen, Niko. *Animal Behavior*. Understanding Science and Nature Series. New York: Time-Life, Inc., 1965.

Walker, Lewis Wayne. *The Book of Owls*. New York: Knopf, 1974.